DID YOU KNOW? BUGS

20 THINGS YOU DIDN'T KNOW ABOUT DRAGONFLIES

DOUG BRADLEY

PowerKiDS press

Published in 2027 by The Rosen Publishing Group, Inc.
2544 Clinton Street, Buffalo, NY 14224

Portions of this work were originally authored by Maria Nelson and published as *20 Fun Facts About Dragonflies*. All new material in this edition was authored by Doug Bradley.

Editor: Greg Roza
Book Design: Michael Flynn

Photo Credits: Cover Rudmer Zwerver/Shutterstock.com; series background freshcare/Shutterstock.com; p. 4 saeedahmad123/Shutterstock.com; p. 5 Rudmer Zwerver/Shutterstock.com; p. 6 EarnestTse/Shutterstock.com; p. 7 Angela Gifford/Shutterstock.com; p. 8 ManoStudioArt/Shutterstock.com; p. 9 RaniCreativeimages/Shutterstock.com; p. 10 https://commons.wikimedia.org/wiki/File:Austrophlebia_costalis_(46844186271)_2.jpg; p. 11 Alex Cooper Photography/Shutterstock.com; p. 12 Ekaterina Staromanova/Shutterstock.com; p. 13 Asta yanuar/Shutterstock.com; p. 14 Natthakorn Sanjaikar/Shutterstock.com; p. 15 Martin Fowler/Shutterstock.com; p. 16 Vitalii Hulai/Shutterstock.com; p. 17 Ernie Cooper/Shutterstock.com; p. 18 Eric Lysenko/Shutterstock.com; p. 19 BlueRingMedia/Shutterstock.com; p. 20 Yuriy Balagula/Shutterstock.com; p. 21 Pawich Sattalerd/Shutterstock.com; p. 22 Shairaa/Shutterstock.com; p. 23 Pavlov Sergei888/Shutterstock.com; p. 24 Tony Tilford/Shutterstock.com; p. 26 tolobalaguer.com/Shutterstock.com; p. 27 Danny Ye/Shutterstock.com; p. 29 noomcpk/Shutterstock.com.

Cataloging-in-Publication Data

Names: Bradley, Doug, 1971-.
Title: 20 things you didn't know about dragonflies / Doug Bradley.
Description: Buffalo, New York : PowerKids Press, 2027. | Series: Did you know? bugs | Includes glossary and index.
Identifiers: ISBN 9781499457391 (pbk.) | ISBN 9781499457407 (library bound) | ISBN 9781499457414 (ebook)
Subjects: LCSH: Dragonflies–Juvenile literature. | Insects–Juvenile literature.
Classification: LCC QL520.B656 2027 | DDC 595.733–dc23

Printed in China

CPSIA Compliance Information: Batch #QSPK27. For Further Information contact Rosen Publishing at 1-800-237-9932.

CONTENTS

GRACEFUL FLIERS

Dragonflies are graceful fliers that make their homes near lakes and ponds. But did you know the dragonfly is also a swift and deadly **predator**? From its six legs to the many lenses of its eyes, some of the coolest dragonfly features help it catch **prey**.

There are about 3,000 species, or kinds, of dragonflies. They can be found on every continent except for Antarctica. You might even find them in your backyard!

Dragonflies can be very beautiful bugs!

DRAGON COLORS

FACT 1

DRAGONFLIES ARE COLORFUL CRITTERS.

While some species of dragonfly are more colorful than others, each dragonfly has its own special look. From yellows and reds to browns and blues, dragonflies can show some wild colors. They might look shiny or even be spotted!

This red dragonfly is called a crimson dropwing or crimson marsh glider.

The wings of the phantom flutterer dragonfly are mostly transparent with a small section of color close to its body.

DRAGONFLY WINGS ARE SEE-THROUGH, BUT THEY'RE ALSO RATHER TOUGH.

FACT 2

A dragonfly's delicate-looking wings add to its beauty. **Transparent** as well as colorful, their wings are strong—even if they don't look it. Dragonflies commonly have a **wingspan** between about 0.8 inch (2 cm) and almost 8 inches (20 cm).

SUN'S OUT!

FACT 3

DRAGONFLIES NEED THE SUN TO KEEP WARM.

You'll see the most dragonflies on warm, sunny days—that's what they like best! Dragonflies are ectothermic (ehk-tuh-THUHR-mihk), or cold-blooded. This means that how hot or cold a dragonfly is depends on its surroundings.

A sunny summer day is your best chance to see a dragonfly.

A dragonfly warms up in the heat of the rising sun.

WARM DRAGONFLY WINGS ARE BETTER THAN COLD WINGS.

FACT 4

In the morning, you might see a dragonfly perched on a leaf, soaking up some sun, or beating its wings very quickly to warm up. This is because dragonfly wings work best when they're warm. Dragonflies may not even come out on cool or cloudy days!

WONDER WINGS!

FACT 5 DRAGONFLIES ARE THE FASTEST FLYERS OF ALL THE INSECTS.

Dragonflies can reach speeds of about 35 miles (56 km) per hour. The southern giant darner dragonfly is known as the fastest flying insect. Dragonflies are such good fliers that scientists have built robots modeled to fly like them!

The southern giant darner dragonfly is native to the east coast of Australia.

DRAGONFLIES HAVE SOMETHING IN COMMON WITH HELICOPTERS.

FACT 6

Dragonflies have two sets of wings that they can move separately from each other. This movement allows dragonflies to fly up, down, or backward, and make quick turns. They can also **hover** like a helicopter!

FACT 7

PREDATORS HAVE TO WORK VERY HARD TO CATCH A FLYING DRAGONFLY.

Dragonflies are incredibly **agile** fliers. Some dragonflies can copy the tiny motions of a watchful predator so quickly and exactly that the predator can't even see them move! This hides them from animals looking for movement that could mean a snack.

This frog is hoping for a quick meal. Do you think the dragonfly will escape?

A dragonfly's wings allow it to make quick, sharp turns as it flies at top speeds.

SMALL FLYING PREY HAVE NO CHANCE OF ESCAPE FROM DRAGONFLIES.

FACT 8

Dragonflies are carnivores, or meat eaters. However, the meat they eat isn't chicken or a hamburger! Dragonflies use their legs to catch mosquitoes, gnats, and even dragonflies smaller than themselves. In ponds without big fish, dragonflies are often the top predator!

FACT 9

GLOBE SKIMMER DRAGONFLIES TRAVEL SO FAR EACH YEAR THAT IT TAKES SEVERAL GENERATIONS TO COMPLETE A FULL TRIP.

Many kinds of dragonflies **migrate** to avoid cold weather. Some migrate to places that are good for **mating**. In 2010, a scientist studied the migration of the globe skimmer. These dragonflies fly from India, across the Indian Ocean, to east Africa and back every year.

The round trip from India to Africa and back is about 11,200 miles (18,000 km)! It takes several generations to complete this trip. Each globe skimmer travels about 3,700 miles (6,000 km).

DRAGONFLIES START LIFE SWIMMING, NOT FLYING.

FACT 10

After adult dragonflies mate, females lay hundreds of round eggs near fresh water. Depending on the species, these eggs are about 0.02 to 0.04 inch (0.5 to 1 mm) across. Once they hatch, the young **larvae** swim away in search of food.

This emperor dragonfly is laying eggs on a lily pad in a pond.

FACT 11

YOUNG DRAGONFLIES ARE A LOT LIKE FISH.

Dragonfly larvae have six skinny legs, two big eyes, and a small body. They breathe in water using gills, like a fish does. Gills remove oxygen from the water so the larvae can breathe. By pushing water out through these gills, nymphs can jet through the water!

Dragonfly larvae are also known as nymphs.

This nymph has caught a fish for dinner!

DRAGONFLY NYMPHS ARE FIERCE PREDATORS.

FACT 12

Dragonflies may remain nymphs for several months or more than a year. Nymphs use pinching mouthparts to catch prey, such as other insects and even tiny fish. Nymphs eat lots of prey so they can grow—and eventually get their wings.

GROWING UP

FACT 13

AS DRAGONFLY NYMPHS GROW, THEY SHED THEIR SKIN UP TO 20 TIMES.

Nymphs molt, or shed their skin, as they get bigger. It's during these many molts that the dragonfly's wings first appear. When it's big enough, the nymph crawls out of the water and molts one last time.

This nymph has just molted for the final time. It's no longer a nymph, it's an adult dragonfly.

DRAGONFLY LIFE CYCLE

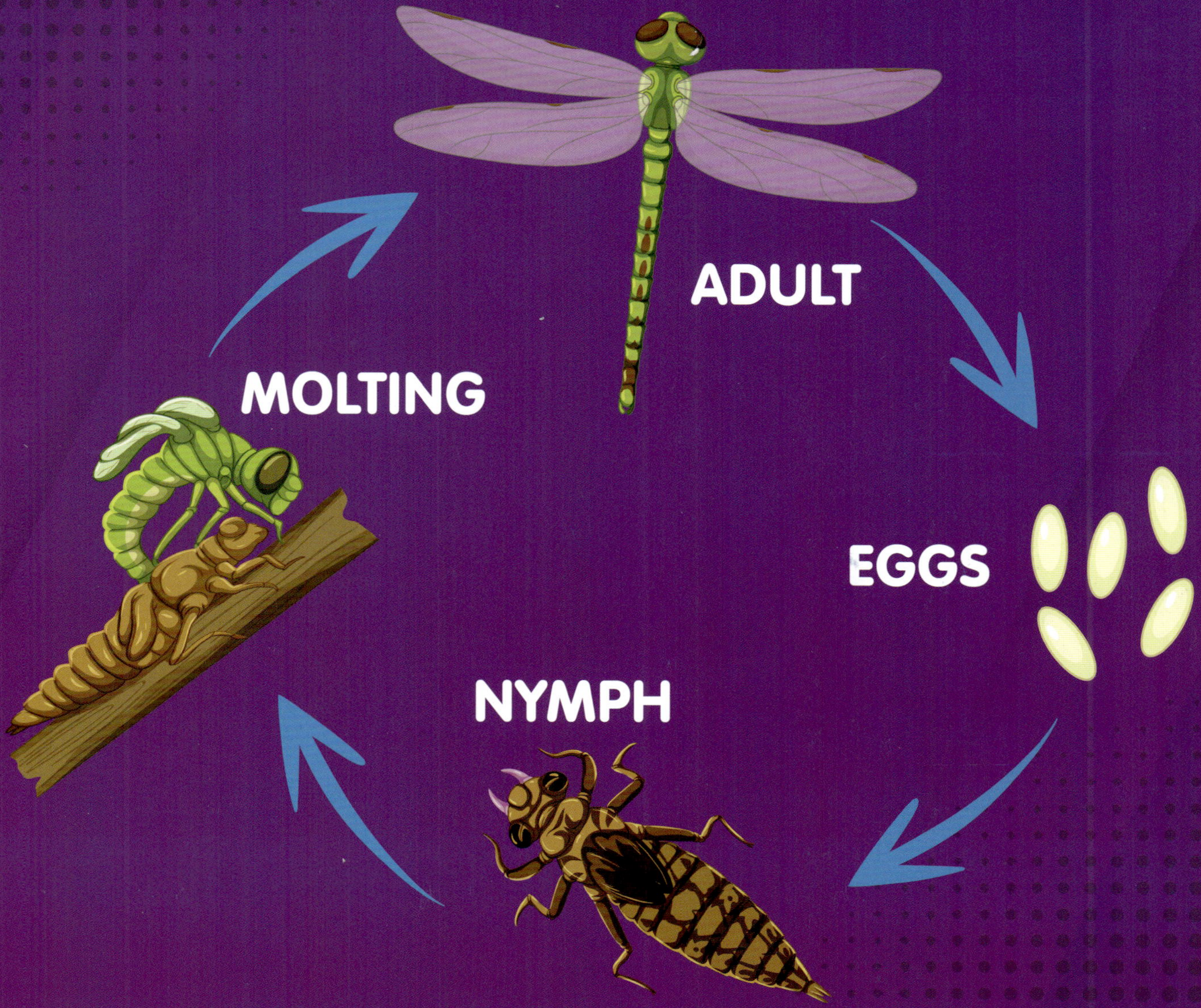

ALL GROWN UP!

FACT 14

NEW ADULT DRAGONFLIES ARE AN EASY MEAL FOR PREDATORS.

After its last molt, the adult dragonfly has to wait for its new **exoskeleton** to harden before it can fly. Predators eat many dragonflies during this time. Ducks, big fish, and frogs are some animals that enjoy a dragonfly meal.

This lucky bird has caught a dragonfly for lunch!

Dragonflies can see almost all the way around their heads.

EACH DRAGONFLY EYE IS MADE UP OF ALMOST 30,000 LENSES.

FACT 15

Dragonflies have sharp eyesight to help them catch prey. Much of their head is taken up with two large compound eyes. Compound eyes have thousands of tiny lenses. About 80 percent of a dragonfly's brain is used for sight!

FACT 16

FEMALE DRAGONFLIES SEARCH FOR A MATE WHO OWNS HIS OWN PROPERTY.

Female dragonflies want to mate with a male that has territory right on a pond or lake. This allows her many good places to lay eggs. Male dragonflies will fight off other males to guard this territory and their mate.

Males select territory that's near fresh water.

THE MORE DRAGONFLIES AN AREA HAS, THE HEALTHIER IT IS!

FACT 17

Dragonflies only live and mate near fresh water that's clean and in areas that don't have a lot of pollution. Pollution affects some species more than others. However, seeing dragonflies is usually a sign that the area is unpolluted.

LOOKALIKES

FACT 18

DAMSELFLIES LOOK A LOT LIKE DRAGONFLIES, BUT THEY'RE NOT THE SAME INSECT.

Dragonflies are part of the scientific group Odonata. These insects have large eyes, two pairs of transparent wings, and a long, thin body. Damselflies are odonates, too! It can be very hard to tell the two apart.

Damselflies rest with their wings together, while dragonflies rest with their wings apart.

DRAGONFLIES AND DAMSELFLIES

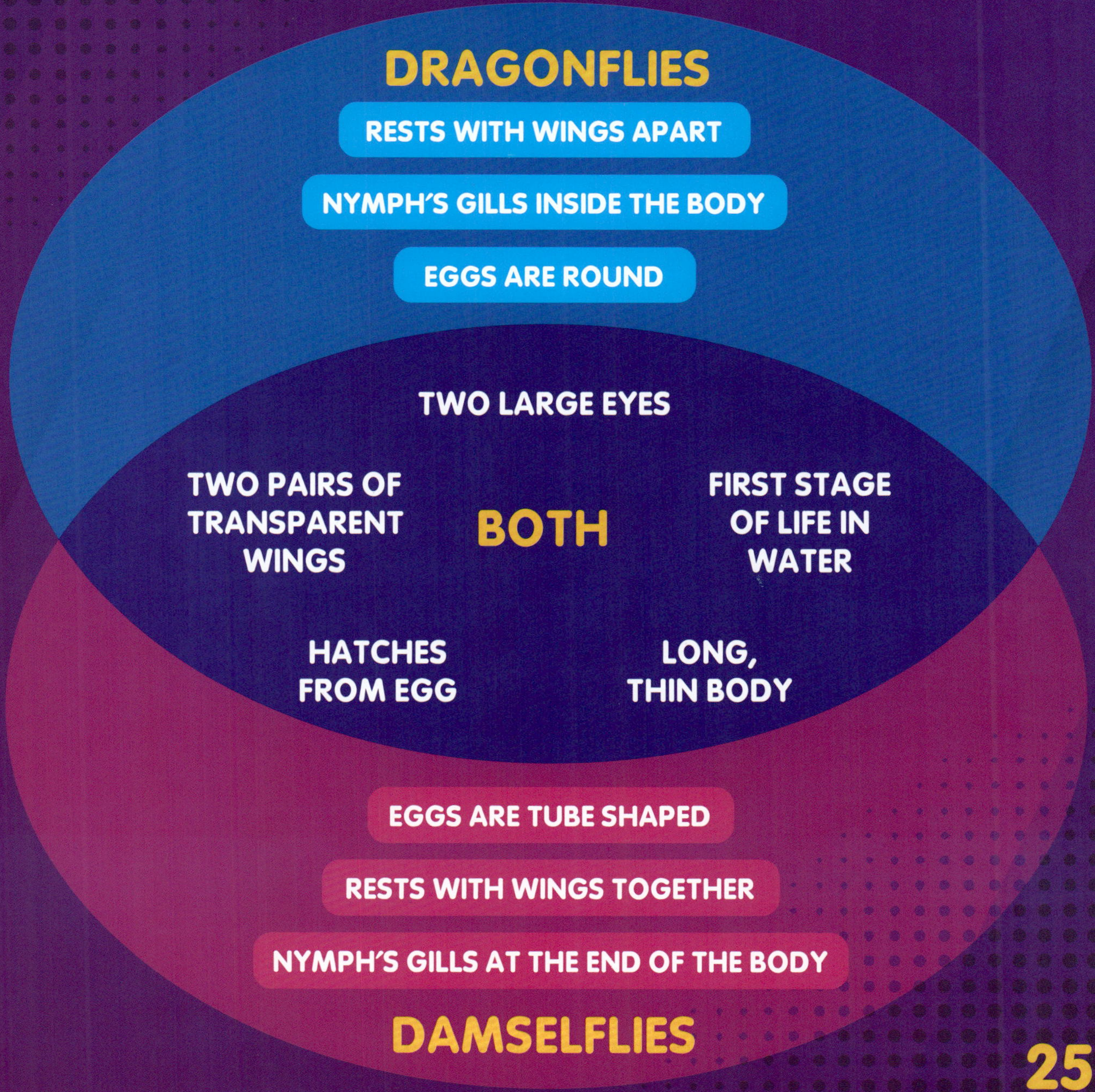

FACT 19

THE FIRST INSECTS RELATED TO DRAGONFLIES WERE ALIVE BEFORE THE DINOSAURS WALKED THE EARTH.

Scientists have found remains, or fossils, from an order of insects called Meganisoptera that are more than 325 million years old. These bugs looked a lot like modern dragonflies. Fossils tell us, however, that they were much, much bigger!

The order Meganisoptera was recently known as Protodonata. "Proto-" is a prefix that means "first" or "original."

This model of an ancient dragonfly shows how big they could grow.

ANCIENT DRAGONFLIES COULD HAVE A WINGSPAN OF UP TO ABOUT 30 INCHES (75 CM)!

FACT 20

The largest known member of the Meganisoptera order is called Meganeuropsis. These monster bugs could grow up to about 19 inches (47 cm) long! That's the largest insect that's ever hovered over the Earth!

ALL AROUND THE WORLD

Dragonflies are cool bugs—from their beautiful colors to their amazing flying skills! Since they're found all over the world—except in Antarctica—people of many countries have stories and ideas about them.

In Japan, dragonflies are more than just something pretty to look at in the garden. They're a sign of power and victory. The Chinese see dragonflies as good luck charms! Native American tribes link dragonflies with speed and happiness. Wherever they're found, dragonflies are sure to brighten anyone's day!

Have you seen dragonflies near where you live?

GLOSSARY

agile: Able to move quickly and easily.

exoskeleton: The hard outer covering of an animal's body.

hover: To float in the air without moving around much.

larvae: Bugs in an early life stage that have a wormlike form.

mate: To come together to make babies. Also, one of two animals that come together to make babies.

migrate: To move from one area to another for feeding or having babies.

predator: An animal that hunts other animals for food.

prey: An animal that is hunted by other animals for food.

transparent: Able to be seen through.

wingspan: The length between the tips of a pair of wings that are stretched out.

FOR MORE INFORMATION

BOOKS

Dijkstra, Klaas-Douwe B., Dr. *Dragonflies and Damselflies of the World: A Guide to Their Diversity.* Princeton, NJ: Princeton University Press, 2025.

Lee, Ashley. *Dragonflies.* Oliver, British Columbia, Canada: Engage Books, 2025.

WEBSITES

Dragonfly Facts
www.dragonflyfacts.com
Read more about dragonflies and see amazing photographs of them.

Dragonfly Facts for Kids
kids.kiddle.co/Dragonfly
Learn even more about dragonflies from this informative website.

INDEX